BIG MACHINES

BIG FARM MACHINES

by Tracy Nelson Maurer

A Crabtree Seedlings Book

TABLE OF CONTENTS

Crabtree Publishing

crabtreebooks.com

093

Big Farm Machines

Big farm machines make big noises.

A tractor chugs across the field.

Tractor tires slurp through mud.

The tractor engine growls as it works.

A mower cuts hay.
Swish-swish!

Harvesters whir up and down the rows.

A combine roars through grain crops.

A sprayer hisses as it sprays crops.

Machines hook together.
Clang! Clang!

JOHN DEERE
CCS Seed Delivery

Crops pour into a truck. Whoosh!

Farm trucks rumble as they haul crops away.

Glossary

crops (KROPS): Crops are plants, such as wheat or potatoes, that are usually grown in large amounts.

engine (EN-juhn): An engine uses fuel to create the power to make a machine move.

farm (FARM): A farm is a place where crops are grown or farm animals are raised.

harvesters (HAR-vis-turz): Harvesters are machines that are used to cut and gather crops.

Index

School-to-Home Support for Caregivers and Teachers

This book helps children grow by letting them practice reading. Here are a few guiding questions to help the reader build his or her comprehension skills. Possible answers appear here in red.

Before Reading

- **What do I think this book is about?** *I think this book is about big farm machines. I think this book is about tractors.*
- **What do I want to learn about this topic?** *I want to learn how to drive a farm tractor. I want to learn what's the most important farm machine.*

During Reading

- **I wonder why...** *I wonder why big farm machines make big noises. I wonder why a sprayer sprays farm crops.*
- **What have I learned so far?** *I have learned that you can hook two machines together. I have learned that farmers pour crops into a truck.*

After Reading

- **What details did I learn about this topic?** *I have learned that big farm machines can drive through thick mud. I have learned that harvesters go up and down all the rows.*
- **Read the book again and look for the vocabulary words.** *I see the word **harvesters** on page 12, and the word **crops** on page 15. The other glossary words are found on page 23.*

Crabtree Publishing

crabtreebooks.com 800-387-7650

Print book version produced jointly with Blue Door Education in 2022

Written by: Tracy Nelson Maurer
Production manager: Candice Campbell
Hardcover 978-1-4271-6101-7
Paperback 978-1-4271-6105-5
Printed in the U.S.A./CP072026

Library and Archives Canada Cataloguing in Publication Title: Big farm machines / by Tracy Nelson Maurer. Names: Maurer, Tracy Nelson, 1965- author.
Description: Series statement: Big machines | "A Crabtree seedlings book". | Includes index.
Identifiers: Canadiana (print) 20210186550 | Canadiana (ebook) 20210186569 | ISBN 9781427161017 (hardcover) | ISBN 9781427161055 (softcover) | ISBN 9781427161093 (HTML) | ISBN 9781427161130 (epub) | ISBN 9781427161178 (read-along ebook)
Subjects: LCSH: Agricultural machinery—Juvenile literature.
Classification: LCC S675.25 .M38 2022 | DDC j631.3/7—dc23

Published in Canada
Crabtree Publishing
616 Welland Avenue
St. Catharines, Ontario
L2M 5V6

Published in the United States
Crabtree Publishing
347 Fifth Avenue
Suite 1402-145
New York, NY 10016

Photo credits: Cover photo and page 8-9 ©Shutterstock.com/NanoStock; page 2-3 ©Shutterstock.com/smereka; page 4-5 ©Shutterstock.com/oticki; page 6-7 ©Shutterstock.com/Jurgis Mankauskas; page 10-11 ©Shutterstock.com/Bonita R. Cheshier; page 12-13 ©Shutterstock.com/Kharlanov Evgeny; page 14-15 ©Shutterstock.com/Lyudvig Aristarhovich; page 16-17 ©Shutterstock.com/gubernat. page 18-19 ©Shutterstock.com/jetcityimage. page 20-21 ©Shutterstock.com/al7. page 22 ©Shutterstock.com/brians101

Library of Congress Cataloging-in-Publication Data
Names: Maurer, Tracy, 1965- author.
Title: Big farm machines / by Tracy Nelson Maurer.
Description: New York, NY : Crabtree Publishing, [2022] | Series: Big machines - a Crabtree Seedlings book | Includes index.
Identifiers: LCCN 2021015181 (print) | LCCN 2021015182 (ebook) | ISBN 9781427161017 (hardcover) | ISBN 9781427161055 (paperback) | ISBN 9781427161093 (ebook) | ISBN 9781427161130 (epub) | ISBN 9781427161178
Subjects: LCSH: Agricultural machinery--Juvenile literature.
Classification: LCC S675.25 .M38 2022 (print) | LCC S675.25 (ebook) | DDC 631.3--dc23
LC record available at https://lccn.loc.gov/2021015181
LC ebook record available at https://lccn.loc.gov/2021015182